一起向未来

人生问答录

图书在版编目（CIP）数据

一起向未来 人生问答录 / 解聘如著；林帝浣绘 .—广州：广州出版社，2022.6

ISBN 978-7-5462-3462-5

Ⅰ. ①一… Ⅱ. ①解… ②林… Ⅲ. ①人生哲学—通俗读物 Ⅳ. ① B821-49

中国版本图书馆 CIP 数据核字（2022）第 079844 号

一起向未来 人生问答录

Yiqi Xiang Weilai Rensheng Wendalu

出版发行 广州出版社
（地址：广州市天河区天润路 87 号广建大厦 9 楼、10 楼
邮政编码：510635 网址：www.gzcbs.com.cn）
责任编辑 卢嘉茜 伍秀娟 李佰幸
责任校对 李少芳
英文翻译 吕翔桢 刘沛怡
英文审校 侯香勤 王松昌
排版设计 南牧文化
印 刷 广东信源文化科技有限公司
（地址：广州市番禺区大龙街竹山工业路 57 号
邮政编码：511450 电话：020-31035505）
规 格 787 毫米 ×1092 毫米 32 开
印 张 6.5
字 数 80 千
版 次 2022 年 6 月第 1 版
印 次 2022 年 6 月第 1 次
书 号 ISBN 978-7-5462-3462-5
定 价 38.80 元

发行专线：（020） 38903520 38903521

如发现印装质量问题，影响阅读，请与承印厂联系调换

序

Preface

顾明远

每个人都有一个人生，有的人活得充实，有的人活得空虚；有的人活得幸福，有的人活得悲惨；有的人活得轰轰烈烈，有的人活得平平淡淡。每个人生活的环境不同，机遇不同，道路不同，结果就不同。但如何去体验人生，却是由每个人的思想修养、价值观所决定的。有的人认为荣华富贵、纸醉金迷为幸福；有的人则认为服务大众、两袖清风为幸福。人生的价值究竟是什么？读了《一起向未来　人生问答录》，你就会得到一个正确的答案。

Everyone has a lifetime. Some lead a full life, while others live either a null, or a happy, or a miserable one. Some live with vigour and vitality, but others may live in peace and mediocrity. As living environment varies from person to person, opportunities, roads, and results differ accordingly. One's approach to life is shaped by his/her ideology cultivation and sense of values. Some regard happiness as having a lot of money and property, while others derive much pleasure from serving the people and having integrity. But what is the real meaning of life? You will find the answer after reading *Together for a Shared Future—Qs & As*.

我也时常在想，每个人的生活都不相同，如何衡量人生的价值？青年人总爱说，要实现自我价值。但什么是自我价值？我想，价值总是在相互关系上产生的，自我存在是无所谓价值的。在人与人交往中，人生价值就有四种类型：第一种是利人利己的；第二种是只利人不利己的，就是通常说的舍己救人的；第三种是只利己不利人的，就是损人利己的；第四种是既不利人也不利己的。这样分类，一目了然，第三、四种恐怕算不得是人生价值。

I think everyone's life is different. But how can we measure the value of life? Young people tend to say that they must realize their own value. But what is value? I think that value is always generated from the relationship between each other, but self-existence has no value per se. In human communication, there are four types of life values: the first is both self-benifiting and others-benefiting; The second is altruism, generally described as benefiting others at the sacrifice of oneself; The third is only self-benefiting at the cost of others; The fourth is not only detrimental to others but also to oneself. Self evidently, the third and the fourth can not be regarded as the value of life.

我还在想，一个人的人生价值是不是还可以体现在四个正确对待上。即：正确对待自然，知道自然对人类的重要性，爱护自然、亲近自然；正确对待社会，有社会责任心，为社会做出力所能及的贡献；正确对待他人，尊重他人，尊重他人的价值观、风俗习惯，善于与人相处；正确对待自己，正确应对自己遇到的挫折或困难，不因失败而气馁，正确对待自己获得的荣誉，不为荣誉而骄傲自满、忘乎所以。

I feel one's life value can be reflected in the 4 "Treatments" as below:

1.Treat nature correctly

Know the importance of nature to human beings; love nature and get close to nature.

2.Treat society correctly

Have a sense of social responsibility; make contributions to the society as much as possible.

3. Treat others correctly

Respect others and their values and customs; get along well with them.

4.Treat yourself correctly

Appoach to the setbacks and difficulties in the correct way; don't be discouraged by failure. Have a correct attitude to honors won ; don't be arrogant or carried away by honor.

这些都是我常常在想的问题，读了《一起向未来 人生问答录》，更引发我的思考。可惜我没有聘如那样的才华，说不深刻，说不生动。《一起向未来 人生问答录》是用诗一般的语言回答了深邃的人生哲理。这里没有冗长的说教，而是简洁地说出了自己对人生的感悟。他的回答，句句都那么精辟。他不是用长江大河的水来冲刷你的身体，而是用滴滴清泉清润你的心灵。我想，读者读了这本书，一定会有自己的感悟。

What is mentioned above is what I often ponder over. After reading *Together for a Shared Future—Qs & As* , I've begun to think more profoundly. It's a pity I don't have such language gift, and I couldn't express myself so well and so vividly as Pinru does. *Together for a Shared Future—Qs & As* answers these profound philosophical questions about life in poetic language. There is no lengthy preaching here, but a succinct statement of his own experiences about life. His answers are so incisive. He does not flush you with flood of water from the Yangtze, but refreshes your mind with drops of water from a clear spring. I think readers will have their own sentiments after reading this book.

聘如同志要我写几句话，我也就写了上面一点感悟。

Pinru has asked me to write a few words, and the above is what I want to say.

顾明远

于北京求是书屋

（顾明远，时任北京师范大学教育管理学院院长，
中国教育学会会长。）

Gu Mingyuan

In Qiushi House,Beijing

(Note: Gu Mingyuan was Dean of the School of Education Management, Beijing Normal University, and President of the Chinese Education Association.)

一起向未来

比寻找温暖更重要的是
让自己成为一盏灯火

1. 问 大千世界，芸芸众生，人真是太复杂了。你能对人下一个简明扼要的定义吗？

In this vast world, a person is a very sophisticated being. Can you give a concise definition of the human being?

答 人？人是世界上站得最直的动物。

The human being? The human being is the animal that stands straightest on the earth.

2. 问 这就是说，人是最高尚的动物？

That is to say, the human being is the noblest animal?

答 是的，人应该是最高尚的动物。但迄今为止，在人群里，什么动物都有。

Yes, they should be. But until now, there have existed all kinds of animals in the human world.

3. 问 那么，什么是人的本质？

Then, what is the essence of a human being?

答 品格，品格是人的本质。

Character.

扫一扫
每日一课

每日打卡

4. 问 你对人的看法？

How do you view the human being?

答 世界上同类物质中彼此差异最大的，是不同人的心。

What differs most in the human world lies in the heart of different people.

5. 问 你主张怎样做人？

What do you advocate in self-conduct?

答 追求而又尽责，不俗而又合群，自信而又谦逊，自力而又助人。

Ambitious but conscientious, unique but gregarious, confident but modest, self-reliant but helpful.

6. 问 你怎样判断人？

How do you judge people?

答 不凭标签，而是看他有无善意和真诚。

Never judge one by the label, but by his goodwill and sincerity.

扫一扫
每日一课

每日打卡

7. 问 人们常说人分好人和坏人。什么样的人才算是好人呢？
It is often said that people fall into two categories: the good and the bad. What kind of person can be regarded as good?

答 好人应该是有责任心、有是非感和有爱心的人。他们的言行能让别人快乐，能使社会增色。
A good person should be responsible, warmhearted and have a sense of right and wrong; Their words and deeds can make others happy and add beauty to the society.

8. 问 你认为人与人之间至少应做到什么？
What do you think is the basis of human relationships?

答 彼此平等，相互尊重。
Mutual equality and respect.

9. 问 为什么有些人不尊重别人？
Why do some people disrespect others?

答 他们不懂得自尊。
Because they don't have self-respect.

扫一扫
每日一课

每日打卡

10. 问 如何才能建立自尊？
How can we gain self-respect?

答 只有尊重别人，才是尊重自己。
Respect is gained by respecting.

11. 问 人的自尊心从何而来？
Where does one's self-esteem come from?

答 羞耻之心。
Sense of shame.

12. 问 我们应该待人以诚吗？
Should one treat others with sincerity?

答 对真诚者不真，那是罪过；对虚伪者赤诚，那是幼稚。
It is sinful to be untrue to the sincere; it is naive to be sincere to the hypocritical.

扫一扫
每日一课

每日打卡

13. 问 人与人之间的距离越近越好吗？

Is the closer the distance between people the better?

答 事实上，常常是身近而心远，身远而心近。

In fact, it could be physically close but spiritually far; it could be physically far but spiritually close.

14. 问 怎么样才算是有眼光的人？

What is an insightful person?

答 既能把远的拉近了看，又能把近的推远了看。

Those who can see the far closer, and the near farther.

15. 问 何谓聪明？

What is cleverness?

答 不自作聪明。

It is not pretending to be clever.

扫一扫
每日一课

每日打卡

16. 问 何谓天才?
What is genius?

答 滴水穿石。
It is perseverance.

17. 问 何谓美?
What is beauty?

答 使人愉悦,使人感动,使人升华,使人奋进。
A force that delights, moves, sublimates, and elevates people.

18. 问 何谓丑?
What is ugliness?

答 使人恶心，使人消沉，使人堕落。
A force that disgusts, depresses, and degenerates people.

扫一扫
每日一课

每日打卡

19. 问 你认为世界上美多还是丑多？

Do you think beauty outweighs ugliness in this world, or vice versa?

答 我赞同罗丹大师的名言："世间不乏美的存在，缺乏的是人们的发现。"

I am in favor of Auguste Rodin's celebrated dictum: "The world doesn't lack beauty, but what is lacked is the people who can discover beauty."

20. 问 世界上最美的是什么？

What is the most beautiful in the world?

答 高尚的心灵。

A noble soul.

21. 问 你认为什么是最美的语言？

What do you think is the most beautiful language?

答 真诚无私的行动。

Sincere and selfless action.

扫一扫
每日一课

每日打卡

22. 问 你认为人与大自然谁更美？

Which do you think is more beautiful, human or nature?

答 总体之美在大自然，个体之美在人，但最美的还是人。

The grand beauty goes to nature, but individual beauty belongs to human, who is the most beautiful of all.

23. 问 何谓善？

What is kindness?

答 爱心。

Love.

24. 问 何谓恶？

What is malevolence?

答 害人之心。

Harmful intentions.

扫一扫
每日一课

每日打卡

25. 问 善与恶的根源？
What are the roots of kindness and evil?

答 公心生善意，私心生恶念。
A public heart begets goodwill, and a selfish heart begets evil thoughts.

26. 问 何谓高？
How could one see a little further?

答 站在巨人的肩膀上。
By standing on the shoulders of giants.

27. 问 何谓强？
What is powerfulness?

答 战胜自我，超越自我。
To conquer yourself and surpass yourself.

扫一扫
每日一课

每日打卡

28. 问 何谓大？

What is generosity?

答 有容乃大，自尊乃大。

To be tolerant, receptive, and self-esteemed .

29. 问 何谓宽？

What is broadness?

答 能拓乃宽，心宽路宽。

To reclaim is to broaden; a broad mind makes a wide road.

30. 问 你认为什么最难？

What do you think is the most difficult in the world?

答 做一个正直的人最难。

To be an upright man.

扫一扫
每日一课

每日打卡

31. 问 那么，什么最容易？
Then, what is the easiest in the world?

答 混世，混世最容易。
To dawdle.

32. 问 你认为什么最可贵？
What do you think is most valuable?

答 人格和国格。
Character and National dignity.

33. 问 你认为什么最可爱？
What do you think is most adorable?

答 纯真。
Innocence.

扫一扫
每日一课

每日打卡

34. 问 什么性格最可爱？
Which character is most adorable?

答 痴于学，痴于工，痴于情。
Be obsessed with learning, work, and love.

35. 问 你认为什么最可敬？
What do you think is most respectable?

答 一身正气，或者忍辱负重。
Be upright, even under humiliation.

36. 问 你认为什么最可怕？
What do you think is most horrible?

答 愚昧，包括陈腐的观念。
A benighted mind with outworn concepts.

扫一扫
每日一课

每日打卡

37. 问 你认为什么最可悲?
What do you think is most pathetic?

答 浪费人生。
Wasting life.

38. 问 你认为什么最可笑?
What do you think is most laughable?

答 傲慢。
Arrogance.

39. 问 你认为什么最可恨?
What do you think is most despicable?

答 造谣，诬陷。
Rumor, frame up.

扫一扫
每日一课

每日打卡

40. 问 何为君子？何为小人？
What is a gentleman and a villain?

答 君子是有志之士，是有责任感、有道德的人；小人只追逐一己之利。
A gentleman has lofty ideals, a sense of responsibility and morality; a villain only pursues his own benefits.

41. 问 何谓伪君子？
What is a hypocrite?

答 言行不一，甚至人面兽心。
Inconsistent in words and deeds, even inhuman.

42. 问 你认为伟人有什么特点？
What do you think are the characteristics of a great man?

答 有非凡的眼光、博大的胸怀和魄力，又像最平凡的人一样朴实。
One with an extraordinary vision, broad mind and courage, but keeping a low profile.

扫一扫
每日一课

每日打卡

43. 问 那么，庸人的标志呢？

Then, what does a mediocre person look like?

答 目光短浅，随波逐流。

Being short-sighted and trending blindly.

44. 问 人们常说人生无常。应该如何面对人生？

People often say that life is impermanent. How should one face life?

答 人生确实无常，我们只能直面现实。

Life is indeed impermanent. We will have to face the reality.

45. 问 人生是一台戏吗？

Is life a drama?

答 我宁愿把人生比作一段路程：你不能漫无目的，也不能漫不经心。

I would rather compare life to a journey. You cannot be aimless or indifferent.

扫一扫
每日一课

每日打卡

46. 问 人生的价值如何体现？

How is one's value of life embodied?

答 社会和弱者需要你。

The society and the weak want you.

47. 问 什么是人应该追求的目标呢？

What should one pursue?

答 人应该追求比生命更久远的东西。

Things that are longer than life.

48. 问 人应当怎样活着？

How should one live?

答 但丁在《神曲》中说得好：“人不能像走兽那样活着，应该追求知识和美德。”

Just as Dante Alighieri said in "The Divine Comedy": "Man can't live like an animal. He should pursue knowledge and virtue."

扫一扫
每日一课

每日打卡

49. 问 那么，人生的真谛是什么？
Then, what is the true meaning of life?

答 追求真善美，创造真善美，享受真善美。
To pursue, create and enjoy truth, goodness and beauty.

50. 问 人生在世，最不能缺什么？
What is the most indispensable in life?

答 最不能缺德。
Morality.

51. 问 你认为人生有几处极其重要的十字路口？
How many important choices do you think one should make during his lifetime?

答 童年心地的培养，少年志趣的确立，青年伴侣的选择，中年入俗而不染，老年成熟而不僵。
To nurture a kind heart in childhood, to determine an aspiration in juvenility, to choose a right life-long partner in youth, to keep a character of integrity in middle age and to be experienced and adaptive in old age.

扫一扫
每日一课

每日打卡

52. 问 人活在世上，应与时代合拍，还是我行我素？

Should one keep abreast of the times or ignore the advancement of the times and persist in their own way?

答 合群合拍不合污。有主见何必我行我素？

Mix with others but stay undirtied. Being independent but not flow blindly with the tide.

53. 问 人生的两翼是什么？

What are the two wings of life?

答 担当和摆脱。没有担当就没有责任感，没有摆脱就没有自由感。

To take up and to put down. Without taking up, there would be no sense of responsibility; without putting down, there would be no sense of freedom.

54. 问 人生最珍贵的是什么？

What is most precious in life?

答 友谊珍贵，时间珍贵，但最珍贵的是信念、意志和努力，也就是自强不息。

Friendship and time, but the most precious is faith, will and endeavor, i.e.perseverance.

扫一扫
每日一课

每日打卡

55. 问 人生的航程中有哪些暗礁？

What are the hidden reefs in the voyage of life?

答 轻信，自满或自卑，重利轻义。

Gullibility, complacency or inferiority, cupidity.

56. 问 你对困难的理解？

How do you understand difficulty?

答 困难是意志和智慧的磨刀石，是成就和成功的阶梯。

It is the whetstone for will and wisdom, and the ladder for achievement and success.

57. 问 你对痛苦的理解？

How do you understand agony?

答 没有追求就没有痛苦，没有痛苦就没有欢乐。

Without pursuit, there is no agony; without agony, there is no joy.

扫一扫
每日一课

每日打卡

58 问 什么时候人最痛苦？
When do people suffer the most?

答 愈是清醒愈痛苦。
The more sober, the more suffering.

59. 问 人怎样才能不被困难、痛苦和邪恶所压倒？
How can one not be overwhelmed by difficulty, pain or evil?

答 战胜自己的软弱。
By conquering his / her own weakness.

60. 问 你对逆境和挫折的看法？
How do you view adversities and setbacks?

答 逆境通向真理，挫折使人深刻。
Adversities lead to truth, and setbacks deepens thoughts.

扫一扫
每日一课

每日打卡

61. 问 一个人突遭厄运时怎么办?

What should one do when encountering sudden misfortune?

答 噩梦醒来是早晨。只要自己的精神不垮，天就塌不下来。

A brilliant morning always comes after a nightmare. As long as your spirit remains, you will not collapse.

62. 问 如果一个人一辈子顺风顺水，到底是幸运呢，还是不幸?

If one goes all smoothly in life, is it lucky or not?

答 平淡也是一种遗憾吧。

Insipidity could also be a pity.

63. 问 乐观的基本信念是什么?

What is the gist of optimism?

答 只要活着，就有希望；只要努力，就会改变。

Living brings hope; striving leads to success.

扫一扫
每日一课

每日打卡

64. 问 何谓好心态？

What is a good state of mind?

答 拥有良好的适应力、承受力和亲和力。

Good adaptability, endurance and affinity.

65. 问 好心态对人很重要吗？

Is it important to have a good state of mind?

答 好心态堪称人生的忠实伴侣。

Yes. It is a loyal partner of life.

66. 问 你认为成功的必由之路是什么？

What do you think is the way leading to success?

答 正确的选择加百折不挠的奋斗。

The right choice and the indomitable attempts.

扫一扫
每日一课

每日打卡

67. 问 你赞成以成败论英雄吗？

Do you judge heroes by their success?

答 事实上，很多失败者也是真正的英雄。

In fact, many losers are real heroes.

68. 问 什么是人最重要的财富？

What is one's most important wealth?

答 健康的身心，良心与爱心，真诚的朋友，智慧与见识。

Physical and mental health, conscience and love, sincere friends, wisdom and insight.

69. 问 何谓幸福？

What is happiness?

答 觉悟人生，造福他人。

Knowing yourself, and benefiting others.

扫一扫
每日一课

每日打卡

70. 问 你觉得什么是感受幸福的能力？

What do you think is the ability to feel happiness?

答 一个人内心世界的丰富与活跃，就是感受幸福的能力。

If a person has a rich and fulfilling inner world, he must have the ability to perceive happiness.

71. 问 人怎样才能获得幸福？

How can one obtain happiness?

答 贝多芬说过："德行是通向幸福的必由之路。"

Just as Ludwig van Beethoven said, "Virtue is the sure path to real happiness."

72. 问 何谓德行？

What is virtue?

答 善言善行。

Kind words and good deeds.

扫一扫
每日一课

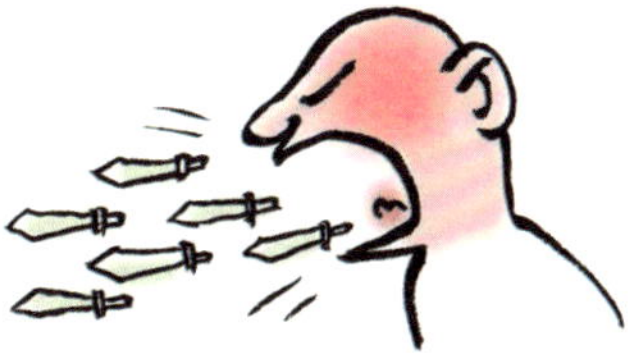

每日打卡

73. 问 你认为做人的起码道德是什么？
What do you think is the bottom line of self-conduct?

答 不当污水，不当锁链。
Never to be nasty, and never to be fetters.

74. 问 何谓时机？
What is timing?

答 稍纵即逝的机会。
Fleeting opportunities.

75. 问 机遇常常属于什么样的人？
Who can get hold of opportunities?

答 积极进取并且善于等待的人。
Those who are well prepared and patient to wait .

扫一扫
每日一课

每日打卡

76. 问 为什么人们会错过机遇？

Why would people miss opportunities?

答 往往是未到站就匆忙下了车。

Because they often get off the train in a hurry before arriving at the terminal.

77. 问 你对命运的看法？

What do you think of fate?

答 谋事在人，成事在天。

Man proposes, God disposes.

78. 问 人生有满分吗？

Is there perfection in life?

答 人生不可能有满分。谁幻想完美，谁就会成为幻想的奴隶。

Life cannot be perfect. Whoever fantasizes about perfection will become a slave to fantasy.

扫一扫
每日一课

每日打卡

79. 问 对一般人来说，什么是完美的？什么是珍贵的？

What is perfection and preciousness for the ordinary person?

答 没有达到的，人们往往以为是完美的；而那失却了的，人们才感到珍贵。

What has not been achieved is thought to be perfect; what has been lost is felt to be precious.

80. 问 人生的顶峰是什么？

What is the peak of life?

答 成熟。

Maturity.

81. 问 何谓成熟？

What is maturity?

答 懂得珍惜。

Cherishing.

扫一扫
每日一课

每日打卡

82. 问 人世间，什么力量最伟大？

What is the greatest power in the world?

答 爱，爱的力量最伟大。大爱无疆，真爱无价。

Love. Great love is boundless, and true love is priceless.

83. 问 何谓爱？

What is love?

答 爱，是给予，是奉献。

Devotion and dedication.

84. 问 你对爱情的理解？

How do you understand love?

答 心息相通，相慕相吸；或者说，没有心灵震颤的天然合拍，就谈不上真正的爱情。

Heart-to-heart, mutual admiration and attraction; in other words, there can be no true love without the natural harmony of two hearts.

扫一扫
每日一课

每日打卡

85. 问 你对失恋的理解？

How do you understand a failure in love?

答 失恋，其实大都是值得庆幸的事情，免得双方以后长期痛苦。

In most cases, to have got the "Dear John" letter is something worth celebrating for avoiding lifelong agony.

86. 问 你认为早恋的结果会是怎样的？

What do you think is the fate of puppy love?

答 早摘的果子苦涩的多。

Unripe fruits are mostly bitter ones.

87. 问 你把家比作什么？

What do you compare home to?

答 家好比港湾。一个温馨的家，无疑是洒满阳光的恬静港湾。

A harbor. A cozy home is a peaceful harbor full of sunshine.

扫一扫
每日一课

每日打卡

88. 问 男人对女人的期望是什么？
What is a man's expectation from a woman?

答 为他营造一个温暖的窝。
To build a warm nest for him.

89. 问 女人对男人的期望是什么？
What is a woman's expectation from a man?

答 成为她可依靠的肩膀。
To become a shoulder she can lean upon.

90. 问 男人和女人互相应该怎样看待？
How should men and women view each other?

答 男人应该懂得，女人不是水果和鲜花；女人应该明白，男人不是财产和宠物。
Men should understand that women are neither fruits nor flowers; women should know that men are neither properties nor pets.

扫一扫
每日一课

每日打卡

91. 问 你认为伴侣之间不可缺少的是什么？
What do you think is indispensable between spouses?

答 理解、信任、尊重、深情。双方还需要不断充实和完善自己，以保持吸引力。
Understanding, trust, respect, and a deep love. Both sides also need to constantly enrich and improve themselves in order to remain attractive.

92. 问 何谓缘分？
What is fate?

答 不期而遇。
Encounter by chance.

93. 问 何谓知己？
What is a soulmate?

答 彻底理解，完全信赖。
Thorough understanding and complete trust.

扫一扫
每日一课

每日打卡

94. 问 你对友谊的理解是？

How do you understand friendship?

答 真诚相待、相助，乃至相知。

Mutual sincerity, support, and rapport.

95. 问 爱情和友谊如何区别？

What's the difference between love and friendship?

答 爱情是整个身心的相属，它是专一的；而友谊如手足之情，多多益善。

Love is fully-devoted and single-minded, but friendship is brotherly-related and multi-oriented.

96. 问 那么，世上究竟还有没有永恒的爱？

Then, is there any eternal love in the world?

答 永恒的爱，不是关系的维系，而是感情的不衰。

Eternal love is not the maintaining of a relationship but a flourishing of genuine affection.

扫一扫
每日一课

每日打卡

97. 问 少年是国家的希望所在吗？

Are young people the hope of the country?

答 是的。梁启超先生曾语重心长地大声疾呼："少年智则国智，少年强则国强。"

Yes. Chinese revolutionist Liang Qichao once said, pleasantly and sternly,"A country will be wise and hopeful if the younger generation are wise and hopeful; a country would be strong and powerful if the younger generation are strong and powerful."

98. 问 你认为，少年时期最要把握的是什么？

What do you think is the most important in one's adolescence?

答 健康的兴趣，健康的朋友。

Having good hobbies as well as good friends.

99. 问 少年应该确立怎样的志向？

What kinds of ambitions do you think should be established by teenagers?

答 正直做人，忠诚报国。

To be honest and loyal to one's motherland.

扫一扫
每日一课

每日打卡

100. 问 **你心目中最崇敬的少年英雄是谁？**

Who is your most respected young hero?

答 **古今中外有过无数少年立志并终生奋斗的英雄，但我最崇敬的是毛泽东。他从少年时代起，就将一生献给为人民谋幸福的伟大事业。**

In the whole world, there have been countless young heroes who have been determined to fight all their lives, but the one I admire most is Mao Zedong. Since he was a teenager, he has dedicated his entire life to the great cause of seeking happiness for Chinese people.

101. 问 **少年阶段重点要克服哪些缺点？**

What shortcomings should be overcome in the adolescent stage?

答 **自私、依赖、懒散。**

Selfishness, dependence, and laziness.

102. 问 **你给青年人的忠告？**

What advice would you offer to young people?

答 **幼稚和虚度光阴，均是不可饶恕的罪过。**

Childishness and wasting time are unforgivable sins.

扫一扫
每日一课

每日打卡

103. 问 你认为综合国力的第一要素是什么？
What do you think is the most important comprehensive national power?

答 国民素质。
National quality.

104. 问 那么，良好的国民素质取决于什么？
Then, what does good national quality depend on?

答 取决于良好的教育。
Good education.

105. 问 你如何估量教育的意义？
How do you view the significance of education?

答 我赞同英国大文豪萧伯纳所说："拯救人类的唯一希望在于教育。"
I agree with the British literary giant, George Bernard Shaw, who ever claimed that,"The sole hope of human salvation lies in education."

扫一扫
每日一课

每日打卡

106. 问 **你认为，老师和家长首先应该教什么？孩子们首先应该学什么？**

What do you think teachers and parents should teach first? What should children learn first?

答 **陶行知先生说得好：“千教万教先教做人，千学万学先学做人。”**

As the Chinese educationist Tao Xingzhi says,"The ultimate goal of education is to be a whole person."

107. 问 **什么是孩子的起跑线？**

What is the starting point for children's education?

答 **学做人，才是真正的人生起跑线。**

To teach them good conduct is the very starting point in their life.

108. 问 **学做人，应该从哪里着手呢？**

How shall we acquire good conduct?

答 **培育一个良好的学习习惯，一份不怕吃亏的人生态度，一种不怕吃苦的坚持精神。**

To have a good study habit and to take a right attitude towards personal gains and losses and have perseverance against suffering.

扫一扫
每日一课

每日打卡

109. 问 作为家长，你感受最深的是什么？
What do you feel strikes you the most as a parent?

答 孩子长得太快，自己提高太慢。
My children's development is ahead of my own self-advancement.

110. 问 你对家长们有何建议？
What is your advice to parents?

答 提高和完善自己，做合格的家长。
To be qualified parents, they should improve themselves.

111. 问 对于溺爱孩子的家长，你有何忠告？
What is your advice to parents who spoil their children?

答 请他们不要忘记一个事实：社会不会溺爱任何人的孩子。
Don't forget the fact the society will never spoil anyone's children.

扫一扫
每日一课

每日打卡

112. 问 溺爱孩子的第一效应是什么？

What is the first effect of spoiling a child?

答 连婴儿都会欺负溺爱他的人。

Even the infants will bully those who spoil them.

113. 问 管束过严的害处呢？

What's the harm of being too strict with children?

答 管束过严，等于拔苗助长。这样的孩子，不是悲观胆小，就是阳奉阴违。

It is like helping the shoots grow by pulling them upward. This will make children either chicken-hearted or feigning compliance.

114. 问 家庭教育，怎样才能达到最佳效果？

How can we achieve the best outcome in family education?

答 多给孩子朗读名著，常陪孩子亲近自然。

By reading masterpieces to the children and taking the children out to be close to nature.

扫一扫
每日一课

每日打卡

115. 问 为什么孩子往往不听家长的话？

Why would children often disobey their parents?

答 请家长先从自己查起：是否言教高于身教？是否压服多于说服？是否言教啰嗦重复，缺乏新意，以至无的放矢？

Parents should reflect: whether they are preaching more than practicing, subduing more than persuading, or chattering without novelty or purpose?

116. 问 对产生逆反心理的青少年有没有行之有效的教育方法？

How can we help teenagers with rebellious psychology?

答 疏导。

By persuading.

117. 问 情商和智商，哪个更重要？

Which is more important, EQ or IQ?

答 情商比智商更重要。智商不足，尚能以勤补拙；情商缺失，轻则难以立足于社会，重则不珍惜自己或他人的生命。

EQ is more important than IQ. Insufficient IQ is a weakness that can be overcome by diligence, while lacking EQ will make it difficult for a person to gain a foothold in society, worse still, it will make him unable to value his life or the lives of others.

扫一扫
每日一课

每日打卡

118. 问 什么是一个人最基本的责任心？

What is a person's most basic sense of responsibility?

答 对自己的生命和人生价值负责。

Be responsible for your own life and its value.

119. 问 你认为自觉人生应包含哪些要点？

What do you think are the most important qualities involved in a meaningful life?

答 热爱生命，热爱学习，热爱祖国。

Love life, love learning and love our motherland.

120. 问 你对学习的看法？

How do you view learning?

答 只有人类才享有的权利。

It's the privilege enjoyed by human beings.

扫一扫
每日一课

每日打卡

121. 问 一个人应如何用功？
How should one work hard?

答 像钻头一样，钻通为止。
Drill like a drill.

122. 问 你认为人类最有意义的工作是什么？
What do you think is the most meaningful job for people?

答 培育儿童，用美熏陶幼小的心灵。
To nurture children by exerting an edifying influence on their hearts.

123. 问 你对低年级教师的看法？
What do you think of primary school teachers?

答 好的启蒙老师，将使孩子终身受益。
A good first teacher will benefit children throughout their lives.

扫一扫
每日一课

每日打卡

124. 问 对于老师，你有何进言？

What advice do you have for teachers?

答 对孩子要有足够的信心和耐心，绝对不要说“太笨”，永远不要说“太晚”。

Be confident and patient with children. Never say "too stupid" or "too late".

125. 问 你认为孩子们最希望从老师那里得到什么呢？

What do you think children want most from teachers?

答 爱心、理解和公正。

Love, understanding and fairness.

126. 问 什么是我们的根本？

What is our root?

答 父母的养育，家乡的水土。

Parents, hometown.

扫一扫
每日一课

每日打卡

127. 问 父爱的伟大之处？

The greatness of paternal love?

答 向导。

Guarding and guiding.

128. 问 母爱的伟大之处？

The greatness of maternal love?

答 无所不舍，无微不至，小中见大。

Mothers's selflessness, carefulness, and attentiveness.

129. 问 同学之间应该竞争吗？

Should classmates compete with each other?

答 与人互助，与己竞争。

They should help each other and compete with the self.

扫一扫
每日一课

每日打卡

130. 问 原谅别人有什么益处?
What are the benefits of forgiving others?

答 使自己胸襟开阔，使氛围融洽，甚至使过错者自愧。
Making yourself broad-minded, making the atmosphere harmonious, and even making the wrongdoers penitent.

131. 问 说谎就是不诚实么?
Is lying dishonest?

答 善意的谎言并非不诚实。
A white lie is not dishonest.

132. 问 一个人失败了，受挫折了，最怕的是什么?
What is worse than failure and frustration?

答 灰心丧气。
Being disheartened.

扫一扫
每日一课

每日打卡

133. 问 做人是毕生的事业吗？

Is self-cultivating a life long job?

答 是的。做事可以不成，做人不能失败。

Yes. You can fail at work, but you can't fail in life.

134. 问 什么是做人的成功？

What is a successful person like?

答 做到问心无愧。

Being free from compunction.

135. 问 什么是做事的成功？

What is an achievement?

答 经得起时间的检验。

It's one that stands the test of time.

扫一扫
每日一课

每日打卡

136. 问 做小事和干大事的关系？

What is the relationship between doing the small and doing the big?

答 只有肯做小事的人，才有希望干大事。

Only those who are willing to do the small can bid fair to do the big.

137. 问 何谓基础教育？

What is basic education?

答 培养良好习惯。

Nurturing good habits.

138. 问 良好的习惯很重要？

Are good habits critical?

答 它是德行的基础。

They're the cornerstones of virtue.

扫一扫
每日一课

每日打卡

139. 问 性情如何培养，德行如何长进？

How can a person cultivate his character and improve his morality?

答 诸葛亮说：“静以修身，俭以养德。”

As a Chinese ancient strategist Zhuge Liang said: "Quietness benefits self-cultivation; frugality nurtures virtue."

140. 问 你认为青年人和老年人各有何弱点？

What do you think is the weakness of the old and the young respectively?

答 我欣赏欧洲的一句谚语：如果青年人不骄傲，老年人不保守，世界将大大进步。

I like a European proverb which says: If the young is not conceited and the old not conservative, the world will leap forward.

141. 问 你对男子汉的理解？

How do you understand manliness?

答 如果以言行粗暴为男子汉的特征，那是提倡野蛮。我以为，真正的男子汉应该果敢沉着，豁达大度，事业执着，感情热烈。

If we regard rudeness in words and deeds as manliness, we are advocating barbarism. I think manliness embodies being courageous and sober, open-minded and generous, persistent in his career, and passionate in love.

扫一扫
每日一课

每日打卡

142. 问 你对成功的理解?

How do you understand success?

答 得到后人的称赞。

To be admired by the posterity.

143. 问 你对成功要素的理解?

How do you understand the factors contributing to success?

答 一靠刻苦和思考，二靠帮助和机遇。

Diligence, careful thinking , external support, and opportunities.

144. 问 你认为成功的钥匙在哪里?

What do you think is the key to success?

答 归根结底在自己手里。

After all, it is in one's own hands.

扫一扫
每日一课

每日打卡

145. 问 你认为成功者有何共同特点？

What do you think successful people have in common?

答 远见，果敢，坚韧。

Foresight, resolution and persistence.

146. 问 信心的意义？

What is the meaning of faith?

答 成功的第一要素。

It is the essence of success.

147. 问 如何走人生之路？

How to take the road of life?

答 别人可以指点，也可以扶你一程，但路还得靠自己走。

Others can give you a map or a hand, but you must walk on your own feet.

扫一扫
每日一课

每日打卡

148. 问 你对中国青年仿效西方生活方式的看法？
What do you think of Chinese youth imitating the Western lifestyle?

答 我主张学习西方青年的自立精神，同时保留敬爱长辈的东方美德。
I advocate learning the self-reliance from the western youth, while retaining the eastern virtues of respecting and caring for the elders.

149. 问 你对知识的看法？
How do you view knowledge?

答 知识是创造力的基础，想象力和思考则是创造力的灵魂。
Knowledge is the basis of creativity, which originates from imagination and thinking.

150. 问 你认为学者们有何共同特点？
What do you think scholars have in common?

答 专心得入迷，重视零碎时间，酷爱思考。
Being engrossed, attaching importance to piecemeal time and loving thinking.

扫一扫
每日一课

每日打卡

151. 问 你对发现的理解？
How do you view discovery?

答 在不同的层次，有不同境界的发现。
Different levels, different discoveries.

152. 问 你对时空的理解？
How do you view time and space?

答 时空是无限的，这是一个真理；对每个人来说，时空都是极有限的，这更是一个真理。
It is a truth that time and space is unlimited; it is more of a truth that for each individual time and space is extremely limited.

153. 问 理想与今天有什么联系？
What is the connection between aspirations and today?

答 理想，就是由一个个今天筑就的。
Aspirations are to be achieved bit by bit, day by day.

扫一扫
每日一课

每日打卡

154. 问 你认为人应当怎样迎接未来？
How do you think one should meet the future?

答 首先积极利用现在，抓紧今天。
Above all, make good use of the present and seize today.

155. 问 在人生的历程中，你最刻骨铭心的感悟是什么？
What is your biggest insight into life?

答 我发现，选择对人生太重要了。如：选择人生目标，选择伴侣，选择合作者……
It is vital for a person to make the right choices at the right time. Such choices involve your life goal, your life partner, and your collaborators, etc.

156. 问 你主张怎样对待人生？
How should we approach to life?

答 积极、顽强、乐观、潇洒。
Be positive, tenacious, optimistic and debonair.

扫一扫
每日一课

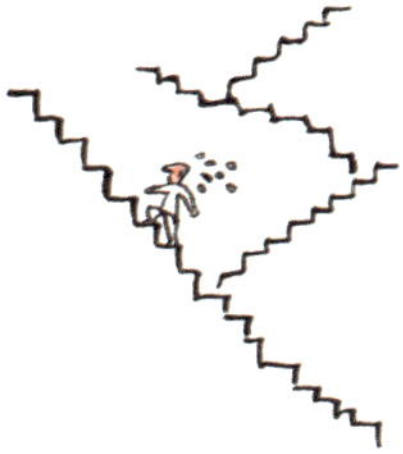

每日打卡

157. 问 上进心的意义？

What is the significance of ambition?

答 没有上进心的人，是真正的未老先衰。

Without ambition, one gets old before one's age.

158. 问 目标如何能达到？

How can one achieve his goals?

答 坚持不懈地走下去。

By perseverance.

159. 问 人都需要别人的帮助吗？

Do people all need help from others?

答 没有别人的帮助，谁也活不了，更谈不上做成什么事。

Without others 'help, one could not survive, let alone accomplish something.

扫一扫
每日一课

每日打卡

160. 问 人应该如何对待环境？

How should one deal with circumstances?

答 适应它，利用它，最终驾驭它。

To adapt to it, to utilize it, and then eventually, to navigate it.

161. 问 人一定要有理想的客观环境才能有所作为吗？

Can one only make a difference in certain circumstances?

答 陶行知先生说："处处是创造之地，天天是创造之时，人人是创造之人。"

As the Chinese educationist Tao Xingzhi puts it, "Everywhere can be the place for creation, every day can be the time for creation, and everyone can be a creator."

162. 问 人怎样才能适应外界的变化？

How can one adapt to changes in the outside world?

答 要有前瞻意识和忧患意识。

To possess a sense of foresight and hardship.

扫一扫
每日一课

每日打卡

163. 问 你认为人应当有怎样的仪表仪态？
What kind of appearance do you think people should have?

答 高贵的头颅，善意的眼神，挺直的腰杆，会心的微笑。
Having a noble head, kind eyes, a straight back, and an understanding smile.

164. 问 你认为人最重要的心理状态应该是怎样的？
What do you think is one's most important mindset?

答 常怀感恩之心。这是人心中不能熄灭的明灯。
Always be grateful. This is the light that cannot be extinguished in the heart of the people.

165. 问 何谓智？
What is wisdom?

答 不为利所诱，不为激所怒，不为危所乱。
Not tempted by profit, not angered by provocation and not disturbed by danger.

扫一扫
每日一课

每日打卡

166. 问 何谓潇洒？

What is debonair?

答 拿得起，放得下；进得去，出得来；尽心尽责去做，对结果、对得失不要太在乎。

The ability to take up and put down or get in and get out. Do something conscientiously, and don't care too much about the results, gains and losses.

167. 问 何谓信念？

What is faith?

答 信念，就是对真理的坚信。

Faith is the firm belief in truth.

168. 问 何谓执著？

What is persistence?

答 认准目标，锲而不舍。

Aim at target and never miss it.

扫一扫
每日一课

每日打卡

169. 问 何谓执拗？

What is stubbornness?

答 十头牛也拉不回头。

It seems as if even ten cows couldn't hold a person back if he is determined to do something.

170. 问 你对忙的看法？

How do you view being busy?

答 蜜蜂忙，苍蝇也忙。

Bees are busy, so are flies.

171. 问 你对钱的理解？

How do you view money?

答 钱太重了，人就轻了。

The more importance you attach to money, the less you will be.

扫一扫
每日一课

每日打卡

172. 问 人为什么会掉进钱眼里？
Why do some people become slaves of money?

答 人变小了，才会掉进钱眼里。
Because they are willing to be controlled by money.

173. 问 你对权力的看法？
How do you view power?

答 权力使正直者多做好事，使卑劣者为所欲为。
Power enables the righteous to do the righteous, and makes the wicked do the wicked.

174. 问 人都是如何被打倒的？
How is one defeated?

答 人都是被自己打倒的。
By himself/herself.

扫一扫
每日一课

每日打卡

175. 问 何谓廉洁？
What is integrity?

答 一身正气，两袖清风。
Being righteousness and clean-handed.

176. 问 人怎样才能抵制诱惑呢？
How can one resist temptation?

答 坚守做人的准则。
Sticking to his principles of life.

177. 问 你对高低贵贱的看法？
How do you view nobleness and lowliness?

答 高，不等于贵；低，不等于贱。
High rank doesn't mean "honorable"; lowly position doesn't indicate "degrading".

扫一扫
每日一课

每日打卡

178. 问 你认为什么最高贵？
What do you think is the most noble?

答 善行最高贵。
Good deeds.

179. 问 何谓天道？
What is the Tao of Heaven?

答 善有善报，恶有恶报。
Good begets good, and evil leads to evil.

180. 问 何谓地道？
What is the Tao of Earth?

答 先有耕耘，才有收获。
No pains, no gains.

扫一扫
每日一课

每日打卡

181. 问 何谓人道?

What is the Tao of Humans?

答 爱护生命，尊重人格。

Treasure life and value dignity.

182. 问 人为什么会生病?

Why could one get sick?

答 社会病了，人的病就多了。

When the society is sick, people in it will suffer hereafter.

183. 问 人为什么会衰老?

Why would one age?

答 心锈损了，人就衰老了。

When the heart rusts, one's life withers.

扫一扫
每日一课

每日打卡

184. 问 人可以青春常在吗？

Can one stay young forever?

答 可以，保持一颗童心。

Yes, by keeping a childlike innocence.

185. 问 你对时间的理解？

How do you understand time?

答 抛弃时间的人，自然也会被时间抛弃。

Those who abandon time will also be abandoned by time.

186. 问 古今中外，人们常常感叹人生苦短。对此你作何感想？

Throughout the ages, at home and abroad, people often lament that life is short. What do you feel about this?

答 俄罗斯伟大作家屠格涅夫曾用一句话点亮了世人的心："我们的生命虽然短暂而且渺小，但伟大的一切都是由人的手创造的。"

Russian great writer Ivan Turgenev once lightened the hearts of the world with a sentence: "Our life is short and limited; but the limitless can be achieved by human beings who have limits."

扫一扫
每日一课

每日打卡

187. 问 你对生命的理解？
How do you understand life?

答 生命在于创造、追求和探索。
Life is to create, to pursue and to discover.

188. 问 人如何能最有效地延长生命？
How can one prolong one's life most effectively?

答 珍惜时间，提高效率。
By cherishing time and improving efficiency.

189. 问 你对死亡的理解？
How do you view death?

答 缤纷的落叶，潇洒飘逸，从枝头回归大地。
It is like golden leaves of a tree gracefully falling to the earth.

扫一扫
每日一课

每日打卡

190. 问 你对人不能长生不老作何感想？

What do you think of the mortality of human beings?

答 幸亏人都是要死的，不然，人的自私性会更可怕，社会也会更难前进。

Fortunately, everyone is mortal. Otherwise, the world will be ruined by people's selfishness, and the society will crank away with difficulty.

191. 问 你对犯了罪的人怎么看？

What do you think of those who have committed crimes?

答 罪犯不一定都是心地坏死的人，他们当中很多人是糊涂人，迷失了，坠入深渊。而能真心悔罪的人，都是可敬的。

Not all of them are completely rotten. Many of them might have got lost, or gone astray, or even fallen into the deep valley. But those who want to gain redemption are to be respected.

192. 问 何谓堕落？

What is depravity?

答 人的堕落，与物的腐败一样。

It is the same as decay of matters.

扫一扫
每日一课

每日打卡

193. 问 有人说，在当今的世界上无所谓是与非、善与恶。你的看法呢？

Some say that being right or wrong, good or evil makes no difference in today's world. What's your opinion?

答 是非不辨、善恶不分，不是极端自私者，就是玩世不恭者。

Only a selfish and cynical soul fails to recognize that difference.

194. 问 人们常说正直的人苦头多。对这个说法，你怎么看？

What do you think of the remark that the honest would suffer more?

答 正直的人很可能会多吃苦头，这就好比责任感强的人往往超负荷。然而，真正懂得人生的人，必定是正直的，责任感强的人。

It is true that those of integrity will bear hardship and those of responsibility will be overloaded. However, those who have a good understanding of life must be those having integrity and responsibility.

195. 问 和善能感化恶人吗？

Can kindness influence the wicked?

答 恶人总是把别人的和善当成软弱可欺。

The wicked always regard the kindness of others as a vulnerability.

扫一扫
每日一课

每日打卡

196. 问 善良人在世俗者眼里的形象是什么?

What is a kind person in the eyes of the secular?

答 傻瓜嘛。

A fool.

197. 问 你对好人常遭非议的现象怎么看?

How do you view the phenomenon that good people are often wronged?

答 这就如同纯净的空气遭到污染一样自然。

It is as natural as fresh air being contaminated.

198. 问 为什么有些人喜欢说闲话?

Why are some people fond of gossips?

答 无聊和缺乏善意。

Because they feel empty and lack empathy.

扫一扫
每日一课

每日打卡

199. 问 你认为哪几种人不可交？
What kind of people do you think are not to be befriended?

答 缺乏真诚者，狂妄自大者，实用主义者，妒嫉心太重者。
The insincere, the self-conceited, the utilitarian and the jealous.

200. 问 你认为狂妄的人有何共性？
What do you think the arrogant have in common?

答 肤浅。
Superficiality.

201. 问 你怎样看待诚信？
How do you view integrity?

答 自尊者诚信。诚信是最好的名片。
Self respect goes hand in hand with integrity, and integrity is a wonderful business card.

扫一扫
每日一课

每日打卡

202. 问 你受了骗怎么办?

What happens when you are deceived?

答 相信恶有恶报。

I believe that evil will be rewarded with evil.

203. 问 如何防止受骗?

How can we avoid being scammed?

答 不贪图便宜。天上永远不会掉下馅饼。

Never be greedy. There's no free lunch in the world.

204. 问 郑板桥说过的“吃亏是福”，如今还适用吗?

Does Chinese artist Zheng Banqiao's "suffering a loss is a blessing" still apply today?

答 一个人肯吃亏，就不会没有朋友，也不会没有前途，更不会没有快乐。

A person who is unwilling to suffer a loss will have no friends or future, let alone happiness.

扫一扫
每日一课

每日打卡

205. 问 人与人怎样才能和睦相处呢？
How can people get along with each other in harmony?

答 少一分自私，多一分和睦。
Be less selfish, be more amicable.

206. 问 微笑的内涵？
What are the implications of a smile?

答 平等，尊重，友善。
They are equality, respect, and friendliness.

207. 问 你对勇敢的理解？
How do you view bravery?

答 战胜自我，不畏人言。
To overcome selfishness and brush aside gossip.

扫一扫
每日一课

每日打卡

208. 问 你对胆怯的理解？
How do you understand timidity?

答 越怕越有鬼。
It's a self-fulfilling prophecy.

209. 问 你对好意的理解？
How do you understand good intentions?

答 成就你的是好意，毁灭你的往往也是好意。
It is what fulfills you, and it is what destroys you sometimes, too.

210. 问 你对拒绝的理解？
How do you understand refusal?

答 要拒绝，最好在开始时。
If you want to refuse something, do it as soon as possible.

扫一扫
每日一课

每日打卡

211. 问 性格能决定命运吗？
Can one's personality decide one's fate?

答 性格至少能影响命运。
At least, one's personality can influence his fate.

212. 问 你主张人内向还是外向？
Which do you advocate, developing inward or outward?

答 内向而不乏朝气和追求，外向而不乏教养和深沉。
Developing inward, you must have vitality and ambition, and while expanding outward, you should be well-educated and self-restrained.

213. 问 你对个性的看法？
How do you view individuality?

答 个性，不是形式上的标新立异，更不是固执和暴躁的同义词；没有正直的独立人格和严肃的独立思考，就没有个性。
Individuality is not a form of unconventionality, nor is it a synonym for stubbornness and irritability; there is no individuality without an honest independent personality and serious independent thinking.

扫一扫
每日一课

每日打卡

214. 问 如何看待个性与群体的关系？

How do you view the relationship between individuality and collectivity?

答 没有个性就没有创造，没有群体就没有壮丽。个性与群体结合了，人类的奇迹就诞生了。

Without individuality there is no creativity; without collectivity there is no splendor. Once they combine, the miracle of humanity is born.

215. 问 个人和社会、民族是怎样的关系？

How do you view the relationships among individuals, the society and the nation?

答 好比水手和船的关系。

It is like the relationship between the sailor and the ship.

216. 问 何谓风度？

What is demeanor?

答 修养的外观。

The appearance of self-cultivation.

扫一扫
每日一课

每日打卡

217. 问 何谓气质?

What is temperament?

答 内在的美。由先天造就,靠后天充实。

The inner beauty, created by nature and enriched by nurture.

218. 问 你对忍耐的看法?

How do you view tolerance?

答 必要的忍耐是坚韧的表现，不必要的忍耐是软弱的同义词。

Necessary tolerance represents tenacity, but unnecessary tolerance equals weakness.

219. 问 何谓软弱?

What is cowardice?

答 绵里无针。

Spinelessness.

扫一扫
每日一课

每日打卡

220. 问 你对嫉妒的看法？

How do you view envy?

答 嫉妒是自卑和自私拌成的毒药，为了伤害别人，却首先毒害了自己。

Envy is a poisonous mixture of inferiority and selfishness. It hurts others and yourself as well.

221. 问 人类最大的祸害是什么？

What is the greatest scourge of human race?

答 贪婪。

Greed.

222. 问 人怎样才能感到知足？

How can one feel contented?

答 懂得珍惜已拥有的。

To cherish what you already have.

扫一扫
每日一课

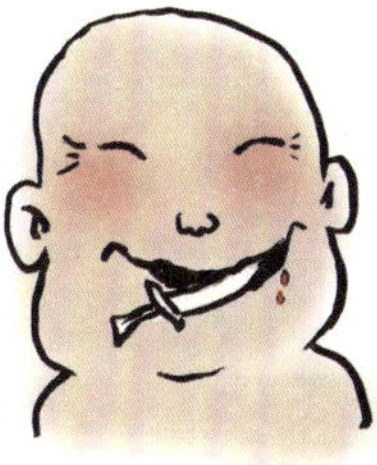

每日打卡

223. 问 知足与自满有何不同？

What is the difference between contentment and complacency?

答 知足，是在索取上无意攀比；自满，是在贡献上不思进取。一般来说，一个人有时确实需要知足，但民族和人类既不可以自满，也不可以知足。

Contentment means no intention in demand; complacency means no intention in contribution. Generally speaking, a person, sometimes, does need contentment but a nation and its people should neither be complacent nor content.

224. 问 你对富有的理解？

How do you understand being wealthy?

答 精神的富有，是真正愉快的富有。

Being spiritually rich is truly wealthy and happy.

225. 问 你认为美好的心灵是如何产生的？

How do you think a kind heart is cultivated?

答 美好的心灵，就是在好人、好作品的熏陶下，或者在苦难的经历中，孕育出的对真善美的热爱和追求。

A kind heart is the pursuit of truth, kindness and beauty, which is bred by being with good people, by reading good works and by experiencing hardships.

扫一扫
每日一课

每日打卡

226. 问 何谓忏悔?

What is repentance?

答 心灵的沐浴。

To purify the soul.

227. 问 何谓宽容?

What is tolerance?

答 一种理解，有时也是一种蔑视。

It's kind of understanding or sometimes, disregard.

228. 问 何谓崇高?

What is loftiness?

答 当你仰望星空，你会觉得天空多么高，也会想到星星看上去多么小，其实多么大。

When you look up in the sky, you will find how boundless the sky is, and how tiny the stars seem to be. But the stars are tremendous indeed.

扫一扫
每日一课

每日打卡

229. 问 何谓幽默？

What is humor?

答 用轻松表达深刻，甚至以笑当哭。

To express profoundness with ease, even to approach to the misery by joking.

230. 问 你对自由的理解？

How do you understand freedom?

答 得到理想的归宿。

An ideal home gained.

231. 问 何谓公正？

What is justice?

答 多劳多得，善有善报。

Work and kindness are rewarded.

扫一扫
每日一课

每日打卡

232. 问 不少人觉得世界不公平，你是怎样认识的？

Many people think the world is not fair. What do you think of it?

答 世界是由美与丑、光明与黑暗、公平与不公平组成的，只要你我都做光明点，光明就不会泯灭，并将一星一点地扩大。

The world is composed of beauty and ugliness, brightness and darkness, fairness and unfairness. If you and I play fair and square, brightness will flare.

233. 问 你对遵纪守法是如何理解的？

How do you understand compliance with laws and regulations?

答 遵纪守法是最可靠的保险单。

It is the most reliable insurance policy.

234. 问 你赞成人们在遭受不公之后实行报复吗？

Do you agree that people should retaliate after they suffer injustice?

答 报复首先会伤及自己。

One will become the first victim of one's retaliation.

扫一扫
每日一课

每日打卡

235. 问 你最爱听的话?
What do you want to hear most?

答 公道话。
Fair remarks.

236. 问 你认为什么人最公道?
Who do you think is most fair?

答 后人。
Posterity.

237. 问 最令你铭记在心的标语口号是?
What is the most impressive slogan for you?

答 "为人民服务。"
"Serve the people."

扫一扫
每日一课

每日打卡

238. 问 你对历史的看法？

How do you view history?

答 要以敬畏之心对待历史。我们今天的言行，都将是明天的历史，它们是无法涂改的。

Treat with awe. Our words and deeds today will be unalterable history tomorrow.

239. 问 你对战争的看法？

How do you view war?

答 侵略战争是最大的反人权。战争和虐待自然，可能是导致人类毁灭的两大祸源。

The invasion wars are the most terrible anti-human activities. War and nature abuse may be the two major sources of human catastrophe.

240. 问 何谓解放思想？

What does "setting the mind free "mean?

答 求实。

It implies "call a spade a spade".

扫一扫
每日一课

每日打卡

241. 问 何谓社会的和谐？

What is a harmonious society?

答 人人都享有尊严。

One where everyone has dignity.

242. 问 你对危机感的看法？

How do you view crisis?

答 有饥饿感的，胃口一定好；有紧迫感的，效率一定高；有危机感的，改革一定快。

The starving must have a good appetite; those under pressure must have a high efficiency; those with a sense of crisis must have a rapid reform.

243. 问 你对中国的富强是否有信心？

Are you confident in China's prosperity?

答 充满信心。改革开放已成滚滚洪流，势不可当；中国不乏有志之士，不乏人才。

Sure, I am. Firstly, the reform and opening up is surging forward irresistibly. Secondly, there is no lack of people of noble aspirations and talents in China.

扫一扫
每日一课

每日打卡

244. 问 何谓诗魂？

What is the soul of poetry?

答 真诚的心声。

Heartfelt aspiration.

245. 问 你在什么时间写诗？

When do you usually write poems?

答 安静的时候，甚至孤独的时候。

In tranquility, even in solitude.

246. 问 你的作品愿给人们带来什么？

What do you wish to bring readers with your works?

答 对人生的珍爱。

Love of life.

扫一扫
每日一课

每日打卡

247. 问 你希望拥有怎样的读者？

What kind of readers do you want to have?

答 真诚的读者。只有真诚的读者才能理解作者的真诚。

Sincere readers. Only sincere readers can understand the author's sincerity.

248. 问 你对文化的理解？

How do you understand culture?

答 个人的品格，民族的灵魂，人类的桥梁。

It is a person's character, a nation's soul, and humanity's bridge.

249. 问 那么，你对艺术的理解呢？

Then, how do you understand art?

答 艺术是从生活中提炼和升华的美和善。

Art is the beauty and the goodness that are distilled out of the real life.

扫一扫
每日一课

每日打卡

250. 问 什么是艺术创作的最高境界？
What is the supreme realm of artistic creation?

答 远离功利，达到忘我。
Putting utilitarianism and ego into oblivion.

251. 问 你最景仰的艺术家是？
Who is the artist you admire most?

答 冼星海和聂耳，他们在民族危亡的关头，为鼓舞民众的斗志创作了永恒的乐章。
Chinese composers Xian Xinghai and Nie Er, who, at the juncture of national peril, created eternal movements that inspired the morale of Chinese people.

252. 问 你认为不朽的文艺作品有何共同特点？
What do you think the enduring artistic works have in common?

答 深刻、创新、震撼心灵。
Profound, innovative, and soul-touching.

扫一扫
每日一课

每日打卡

253. 问 你认为文艺作品是否应该顾及社会效果?
Do you think the creators of literary works should take into account their social effect?

答 严肃正直的艺术家,总愿净化人们的心灵。
A serious artist is always oriented to purify people's hearts.

254. 问 你认为学者与哲人最大的区别是什么?
What do you think is the biggest difference between scholars and philosophers?

答 钻进去的就是学者,钻出来的就是哲人。
Those who get in are scholars, and those who get out are philosophers.

255. 问 人的最佳精神药物是什么?
What is the best psychotropic drug for people?

答 大自然。
Nature.

扫一扫
每日一课

每日打卡

256. 问 你对愉快的理解?
How do you understand happiness?

答 心灵得到充实。
When the heart is fulfilled.

257. 问 你在何种情况下最感惊喜?
When are you pleasantly surprised?

答 发现至真至美时。
When I discover truth and beauty.

258. 问 你对瞬间和永恒的理解?
How do you understand instantaneity and eternity?

答 漫长不等于永恒，美的瞬间却可以成为美的永恒。
Endlessness doesn't mean eternity; a moment of beauty may become eternal.

扫一扫
每日一课

每日打卡

259. 问 那么，什么可以永恒呢？
Then, what is eternal?

答 唯真诚可以永恒。
Only sincerity is eternal.

260. 问 你对自己作何比喻？
How could you describe yourself metaphorically?

答 一个背负十字架的人，一支两头点燃的蜡烛。
A man with a cross on the back; a candle that burns at both ends.

261. 问 你是从郑板桥故里走来的，板桥文化对你有哪些熏陶？
You came from the hometown of Chinese artist Zheng Banqiao. What influence does he have on you?

答 吃亏是福的人生之道，粗茶淡饭的生活之道，梅兰竹菊的修养之道。
The "suffering a loss is a blessing" Tao of life, the minimalist Tao of living, and the noble Tao of self-cultivation.

扫一扫
每日一课

每日打卡

262. 问 你最喜欢什么颜色？
What is your favorite color?

答 一切协调而又富于生气的颜色。如：蔚蓝的大海，碧绿的田野，鲜红的玫瑰，洁白的天鹅……
Anything harmonious and vigorous: the blue ocean, the green field, the red rose and the white swan,etc.

263. 问 那么黑色，你不喜欢黑色吗？
Then, you don't like black?

答 喜欢，黑色的眸子。
I do, the black pupils of the eyes.

264. 问 你最爱听的声音是？
What is your favorite sound?

答 雨打竹林，画眉歌唱。
The sound of rain beating the bamboo forest, and the singing of thrush.

扫一扫
每日一课

每日打卡

265. 问 你最崇尚什么力量？
What power do you admire most?

答 吸引力和创造力。
Attraction and creativity.

266. 问 你对自己的告诫是？
What is your maxim?

答 出淤泥而不染。
Emerging unstained from the filth.

267. 问 你最服谁管？
Who do you follow most?

答 良心。
Conscience.

扫一扫
每日一课

每日打卡

268. 问 你有遗憾吗？
Do you have regrets?

答 人在一生中，总有许多错过和过错，我岂能例外。
In one's lifetime, there are many misses and mistakes, and I'm no exception.

269. 问 你是浪漫主义者吗？
Are you a romanticist?

答 浪漫主义的气质，现实主义的态度，理想主义的追求。
I am one with romanticist making, realistic attitude and idealist pursuit.

270. 问 对于社会和人类，你甘愿做什么？
What would you readily do to the society and humanity?

答 做阳光的一缕，做清泉的一滴，做绿树的一叶。
To be a ray of sunshine, a drop of water, or a leaf of trees.

扫一扫
每日一课

每日打卡

271. 问 当你踏上陌生的土地时，有何感受？

What' s your feeling at the moment you step on the foreign land?

答 天涯处处有芳草。

I feel that I have friends all over the world.

272. 问 你认为见识多的益处是什么？

What do you think is the benefit of having a wider experience?

答 可以减少自大和自卑。

Less superiority or inferiority.

273. 问 你对自己的事业有所成、殊荣有所得作何感想？

What do you feel about the success and recognition you have gained in career?

答 我真切地感悟到，自己是因为得到过无数人的帮助和某些偶然的机遇，才做成了一些事。

I've clearly perceived that I have only achieved something with the help of numerous others and by accidental chances.

扫一扫
每日一课

每日打卡

274. 问 作为炎黄子孙，你最自豪的是什么？
As a descendant of the Yellow Emperor, what are you most proud of?

答 中华民族有着无与伦比的精神财富：博大精深的文化宝藏，百折不挠的民族精神，爱好和平的优良传统。
The Chinese nation possesses unparalleled spiritual wealth: profound cultural treasures, indomitable national spirit and fine peace-loving traditions.

275. 问 你是怎样对待人生的？
What is your attitude towards life?

答 心怀一团火，燃烧工作、燃烧生活。
Work hard, live hard.

276. 问 你的价值观？
What is your sense of value?

答 为祖国、为人民做点有益的事，乃至作出有效的牺牲。
Do something good for my country and people, even at any worthy sacrifice.

扫一扫
每日一课

每日打卡

277. 问 在这个世界上，你心目中最高的是什么？

In this world, what is the most noble thing in your mind?

答 在我的心目中，顶天立地、至高无上的，是天安门广场的人民英雄纪念碑。

In my mind, the Monument to the People's Heroes in Tiananmen Square is the most indomitable and supreme one.

278. 问 你对“读万卷书”有何感想？

What do you feel about the Chinese saying of "reading ten thousand books"?

答 我想起广州少年儿童图书馆有一幅十分醒目的标语：“从小爱阅读，长大最幸福。”

This reminds me of a very eye-catching slogan in the Guangzhou Children's Library: "Lazy youth, lousy age."

279. 问 那么，对“行万里路”，你又有何体会？

Then, what do you feel about "travelling ten thousand miles"?

答 路走多了，视野就开阔了，见识就丰富了，人会变得谦虚起来、勇敢起来、豁达起来，心里既装得下风雨，更充满了阳光。

The more I travel, the wider my horizons are, and the richer my insights will be. Gradually, I become more humble, brave, and open-minded, and my heart is prepared to embrace both storms and sunshine.

扫一扫
每日一课

每日打卡

280. 问 你做好工作靠什么？
How could you make your job well done?

答 笨鸟先飞。
By starting early and working hard.

281. 问 你是乐观主义者？
Are you an optimist?

答 我的心情也有阴晴圆缺，但无论风霜雨雪，我开朗依旧。
My mood is also full of ups and downs, but regardless hardships and setbacks, I'm still cheerful.

282. 问 你觉得自己活得是苦还是甜？
Do you find yourself in bitterness or in sweetness?

答 甜在苦中吧。
Bitter sweetness.

扫一扫
每日一课

每日打卡

283. 问 你是爱国主义者？

Are you a patriot?

答 一切真正自尊的人，岂会无民族自尊心？

How can a person with self-respect have no national pride?

284. 问 你怎样待人处世？

What is your approach to dealing with people?

答 以诚相见，尊人自尊。

With sincerity and mutual respect.

285. 问 你最喜欢和什么样的人交朋友？

What kind of people would you like most to make friends with?

答 青年人。他们可以给我的大脑输氧。

Young people, for they can infuse oxygen to my brain.

扫一扫
每日一课

每日打卡

286. 问 你最大的长处是什么？
What is your greatest strength?

答 心善。
Kindness.

287. 问 你最大的弱点是什么？
What is your vital weakness?

答 心善。
Kindness.

288. 问 你对名和利的态度？
What is your attitude towards fame and gain?

答 我愿享有一切正当的权益，但为了追求理想，我可以舍弃一切。
I wish to enjoy all that I am entitled to, but I would sacrifice all I have for my ideal.

扫一扫
每日一课

每日打卡

289. 问 你最需要的是什么？
What do you need most?

答 理解和时间。
Understanding and time.

290. 问 你忧伤时的表现？
How would you react when you feel sad?

答 沉默。
Silence.

291. 问 你何时流泪？
When do you shed your tears?

答 在美的冲击波下，感动了，或者忏悔了。
When I am moved by beauty, or feel repentant.

扫一扫
每日一课

每日打卡

292. 问 你愤怒时的表现？
How would you react when you are angry?

答 远远地走开。
I'll go away.

293. 问 你怎样对待生活中令你不愉快的事？
How would you deal with an unpleasant experience in your life?

答 把它们从记忆中抹去。
I'll bury them in oblivion.

294. 问 你是否有两重性格？
Do you have double personality?

答 是的。我追求理想，却有时得过且过；我热爱真理，却常常折中退让。
Yes. I aspire for idealism, but sometimes, I muddle along; I love to seek truth, but I often make concession by compromise.

扫一扫
每日一课

每日打卡

295. 问 你如何看待自我？

How do you view yourself?

答 沧海之一粟。

A drop in the ocean.

296. 问 你觉得在日常生活中，什么最难得？

What do you feel are hard to come by in life?

答 难得清静，难得知己，难得糊涂。

It is hard to find peace; it is hard to get a soulmate; and it is hard to know myself.

297. 问 你最大的幸运是什么？

What is your greatest fortune?

答 能到世上走一遭，为祖国和人类创造一点美。

Being able to live in this world and create some beauty for my country and humanity.

扫一扫
每日一课

每日打卡

298. 问 你最大的梦想是什么？
What is your biggest dream?

答 未来的中国人，国民都是君子，领导皆为圣人。
My biggest dream is that in the future of China, all citizens will be gentlemen and all leaders will be sages.

299. 问 你希望向过去索回什么？
What do you wish to reclaim from the past?

答 还我蓝天，还我碧水，还我啼鸟。
The blue sky, the clear water, and the singing birds.

300. 问 那么，你愿意对一百年后的人说一句话吗？
Then, what would you like to say to posterity in the one-hundred-year future?

答 我要用生命刻下一句话：请永远珍惜真诚。
I would like to engrave one sentence with my life: Please always cherish sincerity.

扫一扫
每日一课

每日打卡